1

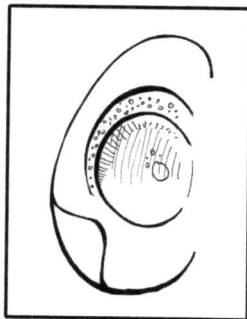

COSMIC AVACADO

Edition 1.52 © 2018 Zarqnon

ISBN-13: 978-0692127575 (Custom Universal)

ISBN-10: 0692127577

Meet the Author and some of his random thoughts...

In 1927, a Belgian Catholic Priest deducted that the universe was birthed at a single point. But what happened prior to that point? We know that our time, our space, and the matter than constructs our universe appears to be self-contained, but what are the plausible constructs prior to that point? We can deduce there has to be a separate time-frame and some state of essence. We suspect there is a "balance" where polarities = 0, yet we also experience "forces" which may imply polarities are skewed/segregated into unequal batches. We know that our universe has stable and consistent parameters that it functions under. Yet, at the very birth of our universe, those parameters appear to be absent.

And what if there was an Architect behind it. Many notions are that a being just "willed it" into existence. But what if that sentient entity designed it with ration and lucidly - constructing it from the very basest modon to the pre-universal fabrics through intricate methodologies. What would be the consequential veracity behind the loop of destruction and rebirth of the fundamental facets that would construct our own spice-thyme-meat-cleaver continuum? What would we expect to see in the enterprise?

Including an Architect in the model comes innately with a rudimentary notion that we are addressing two different time frames: one designated for the universe being created and one that the Architect functions in. But how do these two time-frames function in relation to each other: Are these two time-frames autonomous, is one enveloped in the other, does one branch off the other, or is one or both dependent on the other? We know that the universe being "created" designates a "start" point for the secondary time frame, but does it intuitively consist of an ending point?

My "real" name is JW McLaughlin. My background includes IT stuff, bass player for a metal band, degrees in Educational Research, Business, and Information Systems, and a coffee shop owner. I am an inquisitor of chemistry, biology, physics, music, philosophy, religion, sci-fi and a plethora of other subjects.

3

I am on the autistism spectrum, but did not know this until I was around 40. So I spent most of my life trying to understand my eccentric behaviors and fondness for the peculiar without any "assistance" or "therapy". I described my thought process as "matrix thinking", because everything I had ever learned, observed, and experienced existentially sat on that matrix and was accessible all the time, regardless of context or subject matter - there was no "breaking down of appropriate information due to specific topics".

I work in the field of computers, and one of the things that I have observed is that $\varnothing<>0$. Zero is a cauterized value, and it can skew left, right, up, down, and is parse-able. Zero is a fulcrum, a starting point, a balance. Null is parameter-less. Null, like his sister Infinity, has no boundaries, no possible moves. It is both everything possible and nothing simultaneously. So there is the poly-contextual question: Are we starting from Zero or are we starting from Null?

The reason I am using Zarqnon as the nomenclature for authorship of this book is simply because it amuses me. I like things that are unique and that break the typical - why do it the expected way? If I am violating some preconceived notion of "how it is supposed to be done", then it begs the question "why is it expected to be done that way?" Stretching outside preconceived paradigms gives us the flexibility to experience divergent concepts within paradoxical work frames. I challenge the norm all the time (mostly because I don't understand the norm, and even when I do, I find it "unconvincing" and personally unconventional).

Some extra random thoughts this time from the illustrator...

If you were to walk across a room, you would not have to consciously think to yourself, "Now how do I move my foot? How far? I don't want to fall flat on my face! This is scary." Of course not! All those precise calculations are constantly happening behind the scenes in the majority of your brain's functioning. As an artist, I take this approach to expressing myself through art. The brain is a powerful tool when you say goodbye to anxiety and just let it roam freely.

I discovered this about art not long after seriously injuring my back in a very ill-advised fitness routine with an injured leg. I'll leave the pink dog out of this story and the angry parrot, and how it was all their fault, to concentrate on how it made time to create a lasting positive relationship with art.

I believe that many more people could get joy and accomplishment from art if they could loosen up and not worry so much about what their grey matter is up to. After all, it can be your friend, and it might on occasion stop you from falling over yourself.

I am on the autism spectrum and also legally blind. My blog and coloring books can be located at frankdraws.com

Of Zots and Xoodles:

Theodil Creates a Universe.

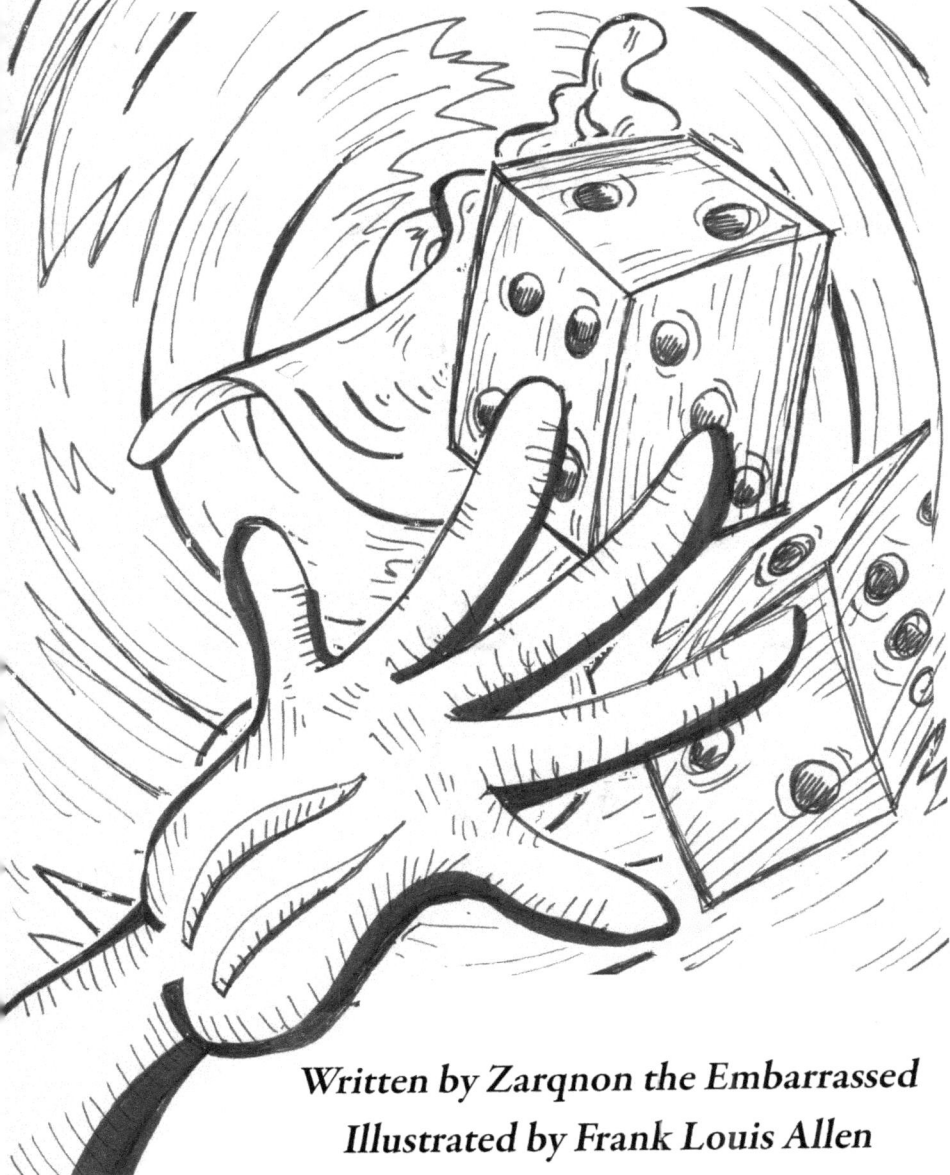

Written by Zarqnon the Embarrassed

Illustrated by Frank Louis Allen

Theodil Creates a Universe.

It would have been sometime after breakfast on a Tuesday, if Tuesdays existed. Theodil walks into the room as the committee stares with a myriad of stares and glares - each pondering their own level of disbelief at what they are about to witness. There, he gives his presentation in that room before rooms, within a space before space, at a time before time: a dimensional playground - a realm of realms where realms are born and realms come to die.

Theodil walks over to the small table. There were those in the crowd who couldn't see Theodil- because people were in front of them, they were sitting the wrong way, they were too busy playing badminton, they had their eyes closed.

> "Maybe he's too big, maybe he's too small, maybe he's hiding behind the wall, maybe he's not there at all!"

He rolls the dice, and dots appear before the crowd.

> "Ooooh" the chamber echoes. "They look like spots. Let's call them Zots. We are so impressed with these dot-like Zots."

And yet before he starts, several in the congregation cry

"This is all wrong! It's taking too long! It's not anything like the song! Sooo, do we really think this is a good idea? Why should it matter? Why are we even here? Has anyone seen my loofah? These chairs stink! They're the wrong color, the wrong size, the wrong shape, and padded completely incorrectly (note: most of them brought their own chair)"

<*One by one, as the Nit-pickers voice their disbelief, their discontent, they gently fade behind the cloak of darkness. But instead of empty chairs, the vestibule re-distributes as though the voices were never there.>

These dot-like Zots are like a spot, yet have not north or south, east or west, up or down or all around and are neither positive or negative, strange or peculiar. These dot-like Zots suspend above his hand as they dance swirling, whirling, churning, in this mystery - absence of space, absence of time - they collide, forming Super-Zots, Mega-Zots, Magma-Zots, and other Zots of Superbious design. And as they dance and assimilate, their properties Pirouette, their attributes allemande, resist and

attract, birthing charge and ricocheting balance left, right, up, down, strange and peculiar.

"Charlatan! Zots should not have knots. Why doth these dot-like Zots birth the wax of gots and not gots. What is wrong with them simply being Zots?!" <*>

A child enters the room and whispers

"Could it be without change, the Zots will never experience fully what can be done through their simplest of design".

"Herumph. I want my Zots to be Zots, not Zots- a-lots. Sustain the status!"<*>

As the voices fade, the Internuncio appears and says "Theodil has created Zots", and then leaves.

As these dot-like Zots of superbious design align and congress, they distort and contort from the dance, the spin, the compression caused by their preponderous aggregation. Theodil pinches the center where the dot-like Zots distort and he pulls. The Zots contend at first, for the Zots are Zots, and their design beckons

them to recede to their first state. But the festering of their innate nature by the imbalance of their amalgamation and their capricious tomfoolery leads them to give way to their fate, their rite of passage, their inevitable progression.

"Ooh, the once dot-like Zots are now like noodles," they exclaim. "We shall call them Xoodles!"

Theodil grins. The once dot-like Zots now noodle-like Xoodles that fare to and fro - some to a "left" and a "right", others to an "up" and a "down" but no further than that. Theodil directs the noodle-like Xoodles to caress and bargain for province. They too, like the dot-like Zots before them, conceded to garnering union. Because of their design, through the synthesized parameters inscribed within the dot-like Zots that gave them birth, the noodle-like Xoodles cultivate idiosyncratic temperament and flare - crimplings, squiglings, folding, looping, expanding beyond the simple to and fro of their youth.

The child takes one of the folded Xoodles and realizes he can shoot Zots all the way from one end to the other at infinite speed. Theodil grins.

The committee marvels as Theodil takes 2 "bowed" Xoodles and touches them together. A particle of light forms at the tangent.

14

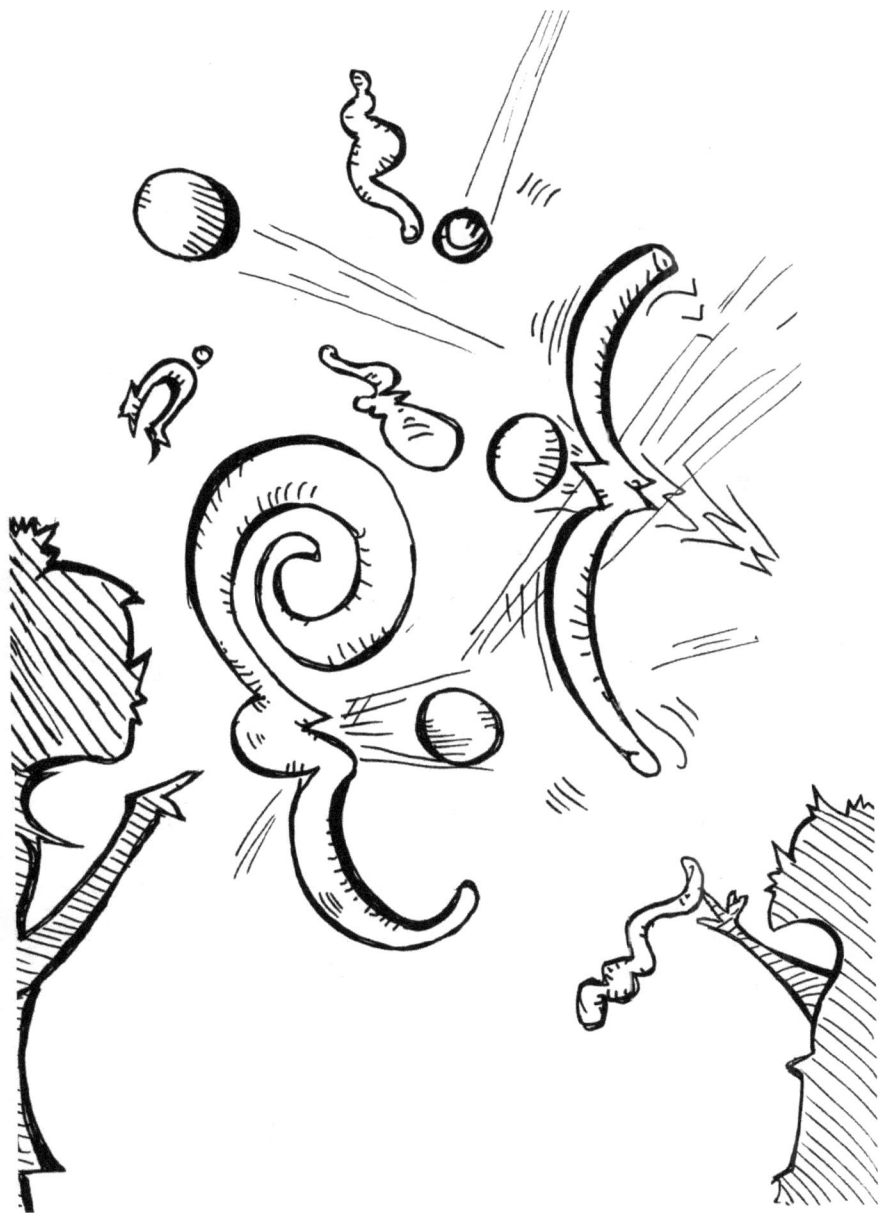

As he scathes them through each other, the particle of light splits into 2 mirror masses which slide along the vortices created by the crossing of the Xoodles. When he pulls the two Xoodles back so that they touch at a single point, the mirror masses meld back to a particle of light.

"Shouldst we such conflict in our conquest for progress? Must we coerce? Could we not simply make Zots as Xoodles to begin with?" "why can't light just be light, and mass be mass? Why complicate the whole thing"?<*>

The child responds with his ponder:

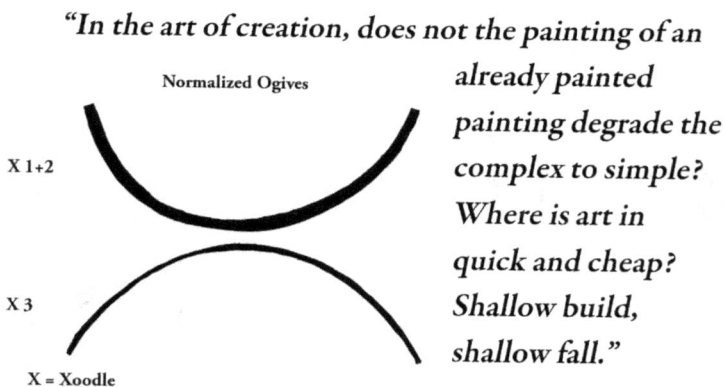

"In the art of creation, does not the painting of an

Normalized Ogives

X 1+2

X 3

X = Xoodle

already painted painting degrade the complex to simple? Where is art in quick and cheap? Shallow build, shallow fall."

Now in his hand, Theodil displays the stack of packets of noodle-like Xoodles. These noodle-like Xoodles, they dangle and tangle, on the brief occasion collide. These collisions create beautiful spark and domain, realms within the realm. And even so, the Beauty unfolds and the beauty fades before the bells toll... the fabrics woven and unravel before a moment pass. Yet even as they rend and tether, they leave behind fragments and parcels. The loss of Zotage as these Xoodles splatter and splinter encourages the Xoodles to return to their dot-like Zot design. Yet to return to the Zots, they can never go back, as Theodil gathers all the fragments and all the splinters and coagulate and cauterize them into a new and improved super-duper Xoodle.

"We can never go back, we can never go back. Is all this destruction categorically necessary? We build, only to destroy? Why don't you just make it perfect to begin with?

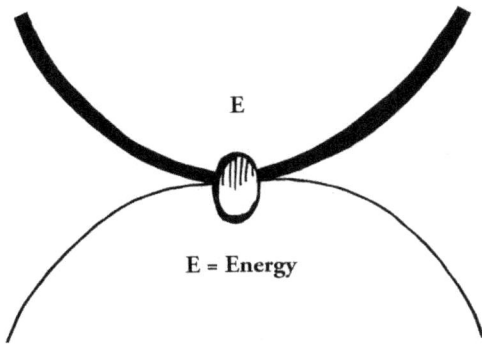

E

E = Energy

The child replies:

"*Maybe, just maybe, the way things need to be when they start isn't how they need to be when they are done. Could it be that one element may need to be strong and unyielding when it begins to crank and chasten the turning and churning of the engine, but if it remains inflexible and intolerant as*

17

the process continues, it hinders the growth of the whole, it becomes a kink within the very machine it intricately assisted in chartering."

Over and over, Theodil repeats the process: taking Zots, stacking them, stretching them into Xoodles, ...

"Why not better be to birth the Xoodles as royalty rather than extract them from peasants?"

M = Matter/Anti-Matter

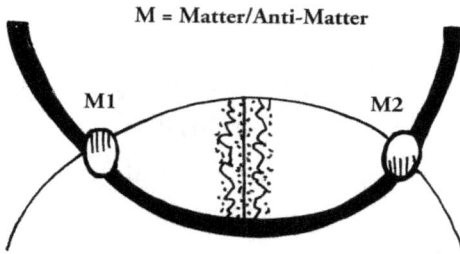

Point where Xoodles break the plane
Fragments and Remnants

"If Theodil has the power to just poof all this shenanigan into existence, why through all this trial and error?"

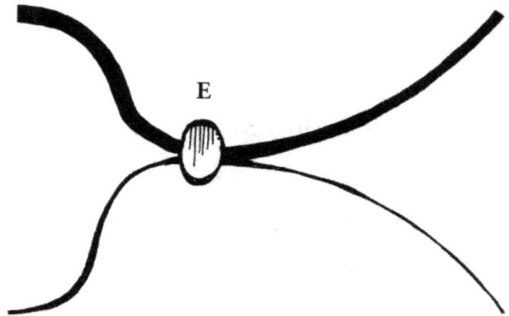

Non-normalized Ogives

"Why so many iterations, when we know you can do it in just one."<*>

"Maybe starting or eventually becoming King or Queen isn't always the optimal and the end goal.

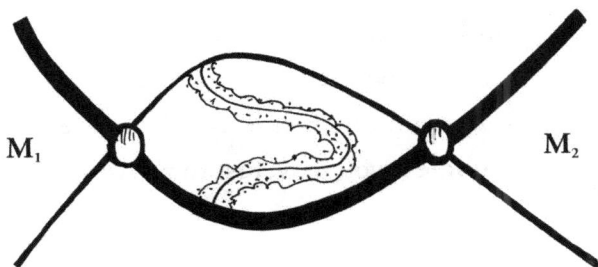

Fragments and Remnants distorted by non-notmalization.

Kings and Queens too easily revert back to pawns in this iteration of the game. Maybe allowing the pawn to pass through trials in the journey allows the pawn to envelope the characteristics needed for the next rite of passage. Otherwise, the vault would be empty and the cave would collapse."

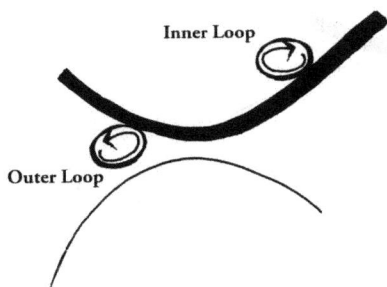

The science enthusiasts drink half of their drink, put the glass down on the table, and walk away.

"Is the opponent always the enemy? How will the pieces progress unless adversity weaves the coat?

The Xoodles continue to collide, coagulate, contort, congregate and jellify into super-Xoodles.

"Remorse we may, let us resend our progress in the midst of this adversity and relent to the sage of the Zot. Preponderance of these Xoodles whelm over our common sensibility. Decimate a Zot or a Xoodle if you may, just to prove you can do it?<*>

"The destruction simply for the sake of destruction would have no glory; it does not prove something. If we cannot see the hand that creates through empirical observation, then why should the hand prove it through destruction. That which breaks

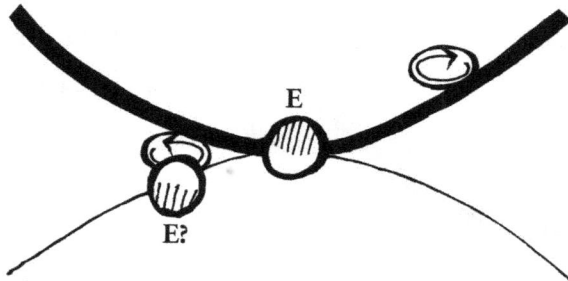

should only break so that it can become part of the journey. Is Theodil doing this to prove he can do it, so that everyone can know he did it, or is there another reason?"

As many Xoodles congregated in super-duper Xoodle packets - each driven through the conjecture of their makeup, they join the chorus of collision and recombine. Theodil wiggles them, which causes them to bend and wave. He blows on them, which freezes and contaminates the parameters of the Xoodles. He folds them, which tucks dimensions and characteristics within the layers. He loops them, creating single dimensional curves and unstable emotions. And he throws them against the wall, shattering them into quantifiable segments with unique spellings.

"The contortion is all we can bare! What did these Xoodles do? Why can't we just use simple noodle-like Xoodles? Why must they be smished and smushed, squashed and squalored together?" <*>

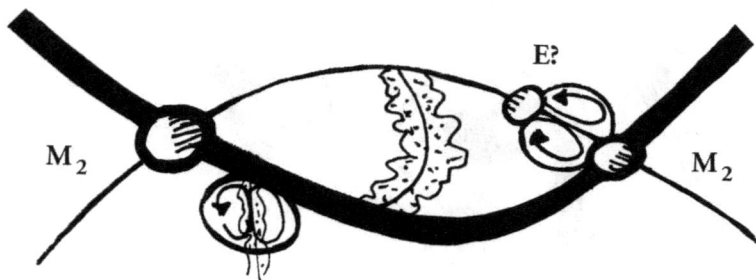

"That which is opposite or different is not always opposing or schismatic. Sometimes it is necessary for the balance. As I pondered previously, I think doing without creating cheapens the process. He is pouring himself into this creation. The evidence of *his heart is in the midst of the process itself. For in the end, the sum of all things is still just the things that sum up the beginning - just the dot- like Zots. But the absolute variance of all things is the experience."*

The Internuncio appears and says "Theodil has created Xoodles", and then leaves.

Over and over, these noodle-like Xoodles collide, weaving fabrics that never taste the platter of time, for the moment they bloom is the moment they dissipate. Comes then the moment trice a Xoodle strikes, a bubble -like grid forms - a new embellishment of breadth and width ever deterring the unraveling as the trimary noodle-like Xoodle binds the weave together, exponentially decelerating the "bubble's" expansion. When the trio noodle-like Xoodles that weave the "bubble's" design strike simultaneously, a giant sphere of light is given birth, but the "bubble" never brings forth gravity, for even with all the various paradigms that are birthed, the fine and balanced distribution of all things with equally distributed pressure encompassing a vortex never allows for unique warps and no grooves in the record for the music to play. There is only "Now" - traveling at the speed of light - a very simple primordial time frame that has no experience. The light without gravity, the plight and the depravity. All things being equal, nothing envelops. To the end that it grows and grows, but never matures, till it pops...

The Internuncio appears and says "Theodil has created Light", and then leaves.

A trio of Simple Xoodles birth a simply singular paradigm fare. This material quickly overcrowds the "bubble" and disallows any progress. But when complex Super Duper Xoodle clusters embark in this union, multiple paradigms of materials are formed as the Super Duper Xoodle splinters into a myriad of intricate fibers: some paradigms embrace the light, others do not.

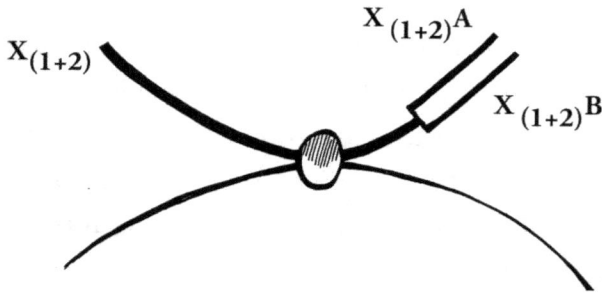

$$X_{(1+2)} \qquad X_{(1+2)}{}^A \qquad X_{(1+2)}{}^B$$

"The Spungic, we shall call it, this avant-garde domain, for it has absorbed all our patience…. And it also kind of looks like Spongebob if Spongebob was a squashed sphere shaped void. Why can't it be oblong, or like a taco?"<*>

Theodil repeats the three strand design, but this time, he pauses until the break of dawn with the third Xoodle so that a planksec passes between when the first two strike and the trimary Xoodle joins in. This allows enough passage so that the light formed by the union can move beyond the cusp of the ultimate urge to return. Again, the striking of the trimary Xoodle greatly slows down the uninhibited expansion, creating a stable parameter for light based matter. This gap between strike one and strike two also creates a domain of time which is correlated with the speed of light. But as of yet, the only time is now. There is no experience, there is only now, and that now spans the distance of progress between strike one and strike two.

Again, the voices of disdain arise:

> "What is all this pandemonius gobbly gook? Are we supposed to be impressed? Is this the Turtle convention? I thought this was the Turtle stacking convention..."<*>

And like the scheme prior, the magnificent Spungic pops, for it appears as the trimary Xoodle strikes the crux, the even distribution of material does not allow for the light to fracture, and if the light does not fracture, the fragments do not matriculate to mass, and the mass cannot aggregate to shape the Spungic through the warping of the fabric.

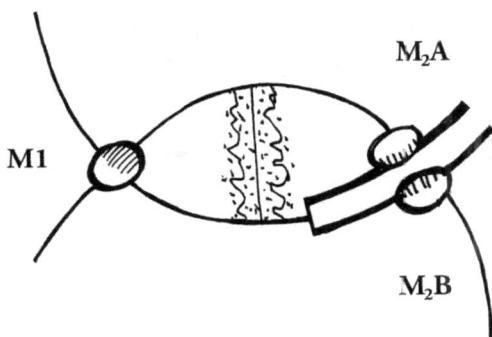

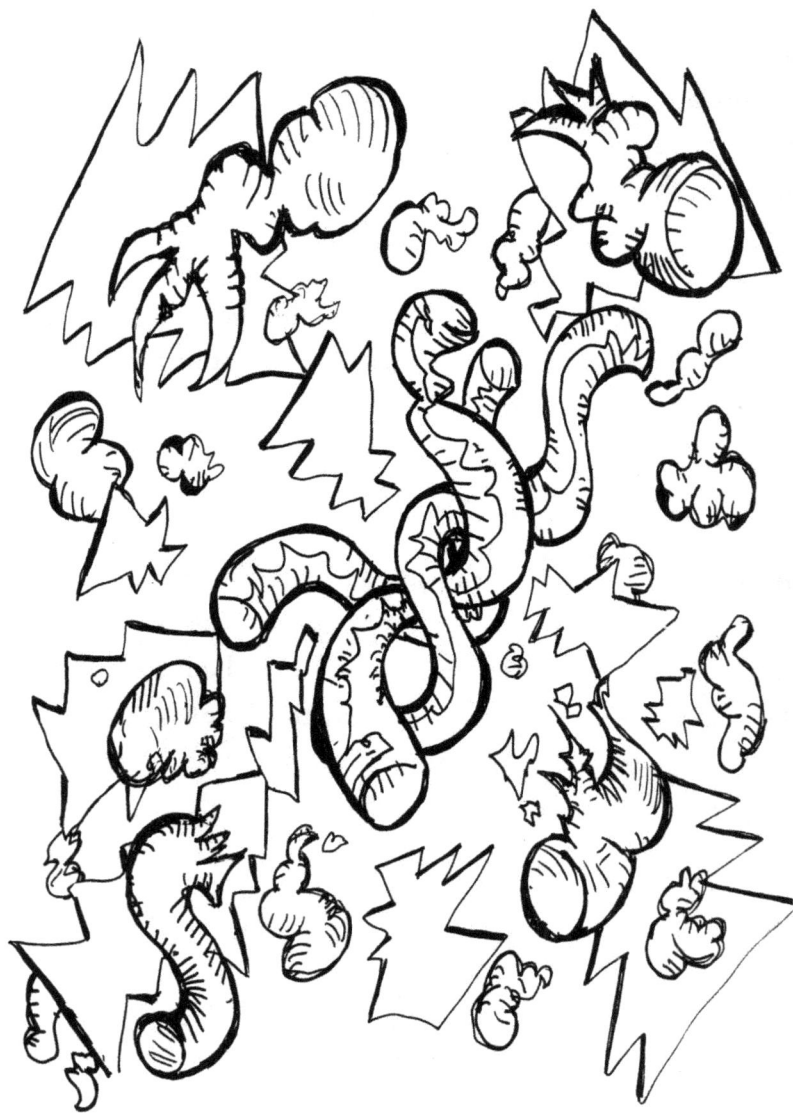

"Can decoherence make us the perfect omelette? I guess that's just another mistake we are going to have to fix. If you will be the spin, I will be the foam. And we shall do the fandango, as the crowd cries out for more..." <*>

Theodil repeats the story of the three noodle-like super Xoodles, Xoodles that swirl, Xoodles that curl. But this iteration, the trimary Xoodle strikes just to the right of the crux. A spin ensued, and a distortion of distribution in light. As the light collides, it gives birth to material, entropy beckons derivation of the light into weaker forms of energy. And as the noodle-like Xoodles hit,

they intertwine, splinter, and they parse into equivalent arrays of mass, including matter and antimatter. Matter and antimatter are pairs in that the sum weight of the arrays on the right are equal the sum mirror weight of the arrays on the left, and the sum weight of the arrays on the up are equal the sum mirror weight of the arrays on the down. Though equal in sum, the arrays were not equal in distribution: the right might have five to the left's three. The up might have one to the down's eight hundred and thirty four. And most significantly, a new paradigm of time is born: The "Now" gives birth to the "Experience of the now": the greater the delay, the greater the experience, the less entanglement, the less experience.

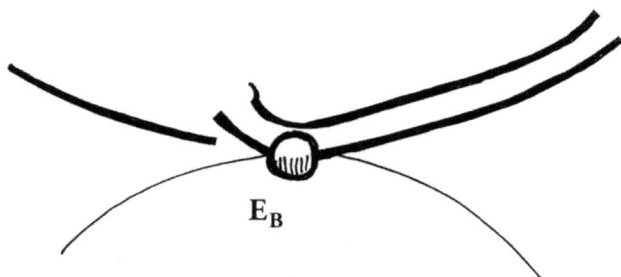

$$E_B$$

Shadow paradigms arise weaving translucent webs where light cannot touch, and the warp that would harken gravity by name will not shake the plane as the shadows pass through. These underlying schemas would encourage gravity without warping and strengthen matter without burning. The entropy that governs the paradigm of light fades in glory when commanding the shadow, for the laws that govern the shadows acknowledge a contrary decree; no light to descend from and no bench mark to be measured. Voices from the committee rise in honest quandary:

"We are noticing in these experiments there are shadow materials that aren't always visible to the eye or interact with gravity like the primary materials"

"If all particles were visible and interacted on the same Schema, would the light be too light? And even with all this light, the overzealous abundance of matter would block the light, so that which needs the light would never see it. When the Xoodles weave the Spungic, everything is just energy. But as energy collides, it parses into various

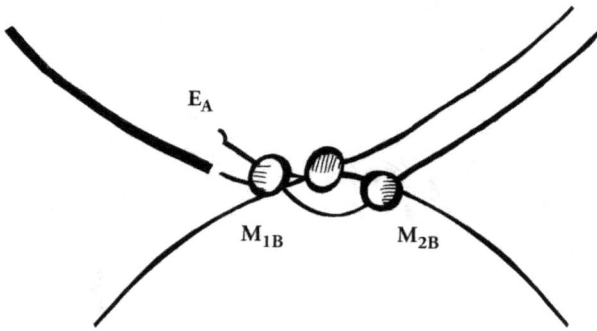

forms of mass, which become vital in the production of gravity. Not every particle will interact with light. Not every material will interact with gravity. As mass increases and accumulates, it will create new mass to the point that it eventually dissipates into new material through explosions and such. Every iteration (explosion) will redistribute gravity and material, allowing for larger and more intricate things..."

"So, now we are having to deal with multiple simultaneous paradigms within a single Spungic? Simple, simple, simple... we want a simple storyline. If the shadow mass is not bound to the same Xoodle segments that construct the light mass, are we to conclude shadow mass bound by the same parameters of speed, experience, and gravity?

"*If all matter had existed at the dawn, the universe would have never had the ability to expand, for the grasp of the entanglement would have been too great. So chaos and entropy are the godparents of progress. Entropy is the metabolic pathway of matter. If all is tangible to light, the mass birthed would prohibit light from grazing in the fields that need light the most. And the heat generated from the entropy of the light would melt the viable universe. The Spungic would dissipate in a blaze of glory. And even if it did not, the eventual growth of mass will crush the Spungic. There must be a proper distribution of energy and mass for the universe to progress. And yet there needs to be another exigence of mass that will govern that which the light cannot.*"

The voices grow ever increasingly impatient:

"Matter, Antimatter, ain't matter, don't matter. If light and matter are of the same paradigm, how can we have matter that attracts other matter but does not reflect light? This shadow matter must have mass, which light does not. And so appears the shadow matter must be intrinsically tying matter together, either through warping the Spungic, or producing an artificial pseudo polar magnetics, or some other existential binding agent".

Most material paradigms become empirically disproportionate because of the offset strike of the trimary Xoodle. The standardized disproportion of the material paradigms that have been disproportionately distributed provides for a systematic gravitational warp within the paradigm: the sum variant of material on the right created an innate struggle for balance in the fundamental osmotic desire for balance. But the material on the right was forever segregated from the material on the left by the parsing trimary Xoodle. The curvature of the Xoodic plane that formed the Spungic would only allow gravity to warp to a given point - a metric determined by the gap between collision one and collision two, at which the end of the warp vortex would contact the Xoodic plane at a relative second point, creating a pin-point funnel between the two points. No matter how hard the material tried, it could not break the barrier of the trimary Xoodle.

Pointing to the disproportion of mass and light, the voices say

"It's not fair! Facets of the Spungic get more than the others" "let the material flow from side to side"<*>

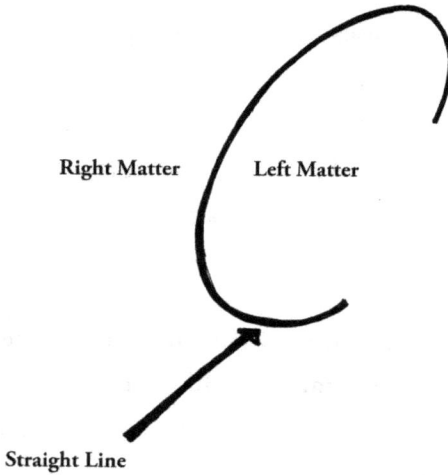

Right Matter   Left Matter

Straight Line

"If the material did flow from side to side, wouldn't that create a variable gravity? Materials would be crushed and fall apart without pity or recourse."

As the scales of time garner providence over experience, it becomes evident that the past is to be inscribed, while the future is not. Then, a Xoodle flickers, and the time engraved on it disappears. And as the Xoodle rests back in its place, the time rewritten did not match the time unwound...

Staring at their clock, another says.

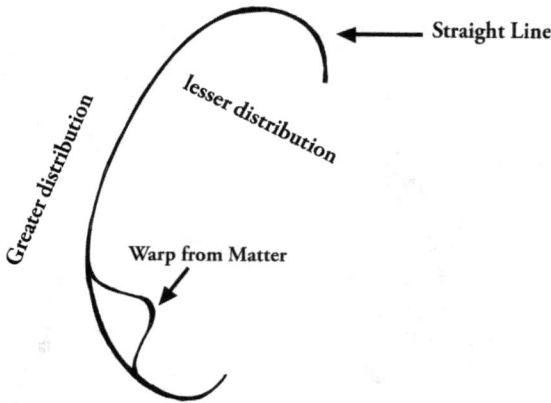

Straight Line

lesser distribution

Greater distribution

Warp from Matter

"A grain of time for a grain of space? Shall our moments and should our Experience be forever governed by the Now?"

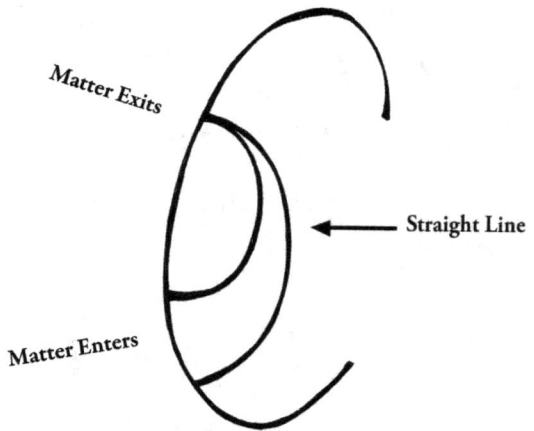

Matter Exits

Straight Line

Matter Enters

"Time that is erased is gone? So if we go back, there is no guarantee the testimony of experience mimics that which was originally there? How can fate weave her dress if the scribe has embedded no set destiny? Fate spins her network of asystematically random sour ramen soup of apostasy."

"The watchmaker has lost his mind. Why should the dance be left to chance?"

"The uncertainty is more than we can bare"

"Could we not, should we not, have time bound in a knot? The past, the future already laid out and neither the past forgot, nor the future an anxious blot, never destined by stain or spot?"

"Why have a past at all? No Future, No Past, only a splotch called Now?"

Light Speed
Past = Now (no experience)

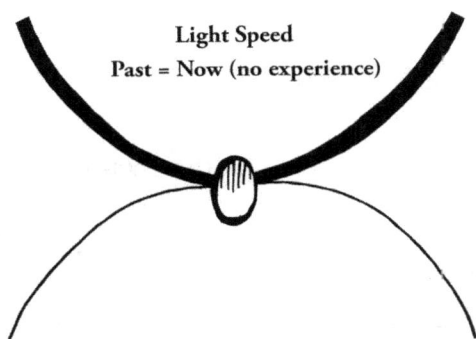

"Why cannot the future pre-exist, and allow our past to unravel, and the now be the point of unjoin? Like a zipper unzipping, as the Now unbinds and fades away. That way we can at least know the future and have our mistakes erased. Why should it be left for us to construct the future?

"Where does accident and random marry? To have a simultaneous construct where all experience was already enveloped in the Now, there would be no room for random events that would give the Spungic a chance for free will."

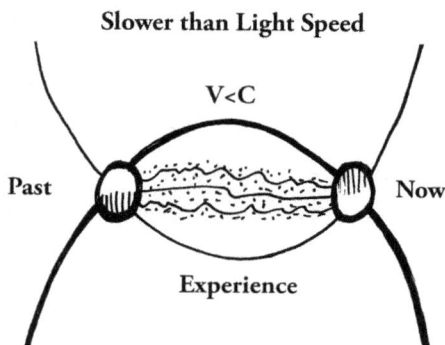

Slower than Light Speed

V<C

Past

Now

Experience

"If there would be no past experience, there would be no point of progress. The present needs to experience the past to have a point to move forward."

"If there was no past but the future, and the point of experience is at the deteriorating end - "Now" would simply be the decaying point of the past. The painting would rot, and in the end, nothing would remain."

"Can infinity exist without time? Even infinity has boundaries in which it functions and thrives."

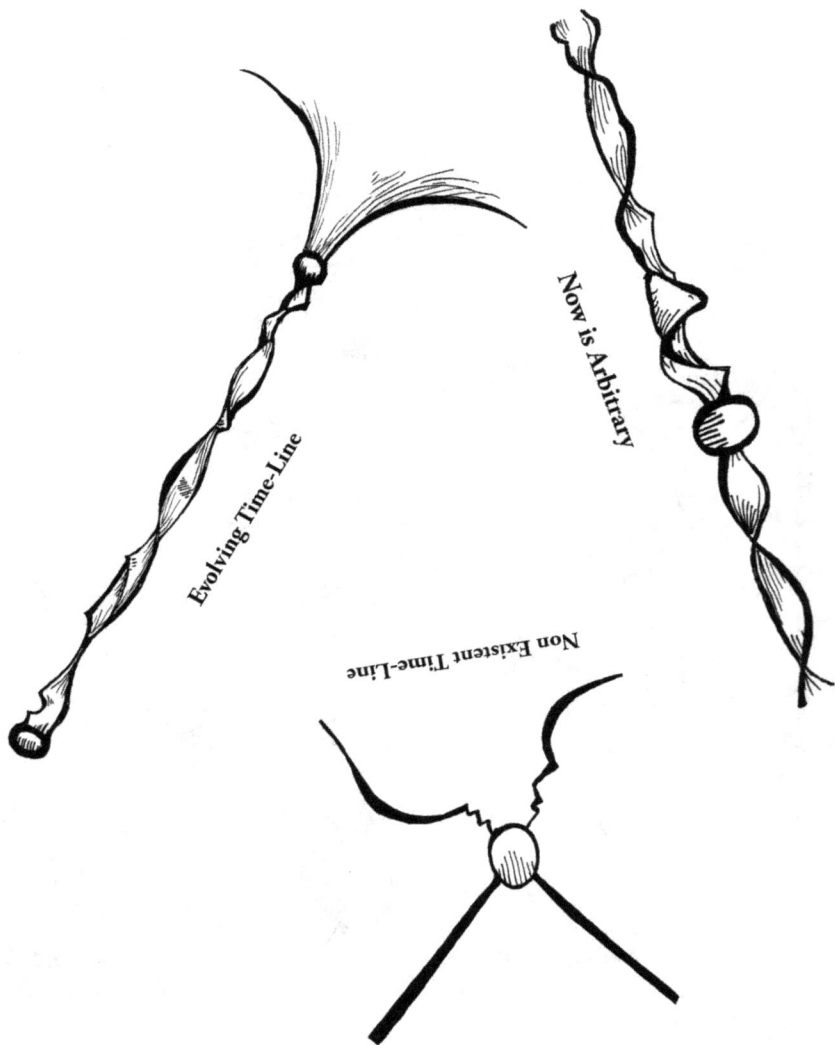

Evolving Time-Line

Now is Arbitrary

Non Existent Time-Line

The Internuncio rings out "Theodil has created time."

Theodil looks at his grand creation. He grabs one of the Xoodles and pulls, and the Spungic collapses into a buzzing, eclectic super dooper Xoodle. Nothing more than a mere whisper and a chorus, Theodil steps back from his work- table.

"You just undid everything you just did? Is this going to be a one-off, or are you going to repeat this whole thing again? If you do, are we required to attend?"

"Haven't we seen this all before? When's the show going to start? My crumpets are burning! We may all be from Copenhagen, yet our worlds are only parallel." the last fading voice calls out· <*>

The Internuncio cries out "the preparatory work is done. The initial stage is over. Now the hysterics can commence".

The child comes and stands next to Theodil.

"Who were these people in the audience? Where did they go?"

Theodil turns and says

"Phantoms: voices to come, voices that will complain in disdain, judge and fudge, ignore and implore, deny and cry, criticize and close their eyes."

Theodil grabs the dice and hands the child a set of dice. Theodil throws the dice upon the floor, and the child does the same.

And the Internuncio, with a voice raised, says......

In the beginning...